TOO CLOSE FOR COMFORT

To all who may assent

Orwell F. Song

Orville K. Long

TOO CLOSE FOR COMFORT

Short Stories With A Common Thread

Orville K. Long

Elderberry Press, LLC

Elderberry Press, LLC
1393 Old Homestead Drive, Second floor
Oakland, Oregon 97462—9506.
TEL/FAX: 541.459.6043
www.elderberrypress.com

Elderberry Press books are available from your favorite
bookstore, amazon.com, or from our 24 hour order line: 1.800.431.1579

Library of Congress Control Number: 2003111273
Publisher's Catalog—in—Publication Data
Too Close For Comfort/Orville K. Long
ISBN 1-930859-76-7
1. Military History.
2. Memoir.
3. Airplanes.
4. Air Corps.
5. Flight.
I. Title

This book was written, printed and bound in the United States of America.

ACKNOWLEDGEMENTS

It is my intention here to give credit where credit is due. I thank the Lord for seeing me through so many close calls. Without His help, a slight misstep anywhere in the following accounts would have meant a sure and timely death. It is also my belief that God helps those who help themselves. These stories speak for themselves.

DISCLAIMER

A. It is to be noted that I am not in the military service (Air Force) and that any of my writings or speaking engagements do not reflect the views of and are not endorsed by the Air Force.

B. It is also noted that I am not employed by the FAA and that my writing or speaking engagements do not reflect the views of and are not endorsed by the FAA.

C. I also wish to thank the Air and Space Museum at Dayton, Ohio for permission to use their airplane pictures.

Orville K. Long
USAF (retired)
FAA (retired)

INTRODUCTION

The author was born on a dairy farm in Minnesota on October 19, 1919.. He attended school in an eight grade, one room country school house.

Mr. Long took an early interest in aviation by watching Northwest Airlines planes fly over his father's farm on their approach to a landing at the St. Paul Municipal Airport.

In his words, "My close encounter with aviation was in my home town, Forest Lake, Minnesota." After a visit with schoolmates there, I departed town on Route 61S. It was then that I became aware of a yellow colored biplane. Checking further, I noticed a sign that read "Fly for $5.00 and 15 minutes." Not to be deterred, I doubled back and parked off the runway near the plane. Joining the spectators, I found my only source of revenue, a five dollar bill. At that time the pilot was loading a passenger in the rear seat. I watched the plane take off and then return in fifteen minutes. As there was no one waiting to fly next, I stepped forward with my hot five dollar bill.

The pilot briefed me that he would take off, circle the lakes and town and land after fifteen minutes. We then took off and leveled off at fifteen hundred feet. The first thing I noticed was the series of three lakes that comprise Forest Lake and the village. What I saw was unbelievable, a forested area around the lakes, the farmers' fields of all different sizes and shapes, the roads leading into town and the myriad of houses of all shapes and colors with their Chevrolets and Fords parked in front. What I saw below me could only be called a Rembrandt, extremely beautiful and fascinating. Then, too quickly, the pilot cut the throttle and landed. If it were not for that five dollar bill in my pocket, I probably would still be on the farm or in the army or navy.

PREFACE

Since 1942 when I was a private in the Air Corp and witnessed my first airplane crash, I often thought that a book of stories should be written showing how close some pilots or passengers have come to giving up their lives.

Enlisting in the Air Corp in 1941, I was sent to airplane mechanic school at Chanute Field, Illinois. Becoming a pilot in 1943 I was sent to many pilot schools. After graduating as a pilot, I specialized in multi-engine aircraft. A few of the major aircraft I flew were B-24, B-29, B-36, C-130 and the B-47.

In the aforementioned case, while a P-43 fighter plane was landing (appendix p.1) his brakes locked and the plane tipped over headfirst, winding up on its back. The canopy was crushed and the pilot was still inside. Gasoline was everywhere. With the help of fire truck members, we lifted the tail section of the airplane high enough to get the pilot out. It was then we became aware that he was unhurt and ready for a smoke. In fact, he had a cigarette in his mouth while waiting to be rescued. These then are accounts of death-defying experiences and close calls.

TABLE OF CONTENTS

Chapter 1

PT-22 AIRCRAFT

During my early training in the PT-22 at Ryan Field in Hemet, California, my instructor and I flew out to an assigned work area to practice loops, among other things. My instructor had me "follow through" on the use of the flight control "stick" as he was going to demonstrate the first loop. As we dove the plane to pick up a speed of 115 miles per hour we found ourselves with an imminent midair collision (see Appendix p.2). We were about to dive straight into another PT-22. It was too late to pull up or turn; we were committed to a crash.

The first thing I did was to push the control stick full forward, as the British would say "with great vigor." I knew for sure that the other plane would crash head on into us. As we plummeted down in front of this other aircraft, I swear I could read the serial number on the engine. In the next microsecond, it seemed that the other aircraft would miss both my instructor and myself, but that he would surely decapitate our plane's tail assembly. Fortunately it didn't happen. We had escaped to fly another day.

After completing other maneuvers, including a properly executed loop, we landed at the completion of our training period but it was a flight that shall never be forgotten. After landing, but before we left, the instructor said I would not have any problem getting my "wings." You see, as a student I had acted alone in dodging the other aircraft. There had been no time to inform the instructor of the oncoming plane.

Chapter 2

THROTTLE STUCK IN IDLE

After finishing Primary Flying School I moved on to Gardner Field near Taft, California. This was a sixty-hour flight course in the BT-13A, commonly called the 'Vibrating Vulture.' The BT-13A had a 450 horsepower engine. This training would include formation flying, cross-country flying and night flying.

While on my third solo night flight I was assigned to the radio range to fly in a holding pattern altitude of 8,000 feet and to the northwest leg of the radio range. After practicing all flight maneuvers I was called back to the Field, but first, descending in thousand foot increments as the other planes were called in from the holding pattern. When all aircraft below me were cleared, I descended to the four thousand foot level. While descending in the "stack" I had retarded the throttle to the idle position.

As I reached four thousand feet I pushed, or tried to push, the throttle forward to full power to maintain altitude. For some unknown reason the throttle stuck in the idle position and would not move forward. Here I was without power, with aircraft below me, and I would be forced to descend.

Things being what they were, I declared an Emergency to the tower. They immediately cleared planes below and then cleared me to the Field, which was about twenty miles distant. I headed in that direction, wondering what could happen next. Luckily, I reached the Field and set myself up for a landing. By making S turns, I lowered the plane to pattern altitude and landed on the runway.

The tower directed me to turn off at the diagonal taxiway where my instructor was waiting for me. He climbed on the wing and reaching in, he tried forcing the throttle forward but to no avail. He had me call the tower to be towed in.

When checking with maintenance the next day, I found they had not discovered why the throttle had "jammed." All pulleys and cables appeared normal. Due to my heavy flying schedule, this particular plane never appeared again on my schedule.

Chapter 3

FROZEN ON THE FLIGHT CONTROLS

During the last week or two of Advanced flying at Stockton Field, California, I was asked to fly as a safety observer for another pilot. This was a BT-14 (see App.p.3) used for instrument rides. The pilot in the front seat was to accomplish two hours of "hooded flight." This meant that a curtain was placed over his head and windshield so that he had no outside reference to the horizon.

The flight went quite well until the pilot felt that he needed a rest period. He pulled back the canopy to get some fresh air and rested for a few minutes. Instead of resuming simulated instrument training, he decided to practice "slow rolls." After failing to talk him out of it, because this was an "instrument only" airplane, he went ahead to perform this maneuver. This consists of rotating the aircraft about its longitudinal axis. He "chickened out" after one-third of the maneuver. Not deterred by his bad performance on the "roll" he proceeded to do a second "slow roll." On this one he got half-way through the "roll" and again "chickened out." At this point he fought desperately to bring the aircraft under control. He finally did.

Despite my admonishment to not try any more "slow rolls," he proceeded to go for this the third time. We were at 8,500 feet. As you might have guessed, when we were upside down he froze on the controls, full right rudder, full left aileron and full up elevator. We were then in a high-speed spin to the right. The only thing outside of the aircraft that I could see was a dollar-size hole straight ahead that showed a farmer's corner fence posts rotating rapidly. I understood very quickly that this was going to be our point of impact, that is, unless something was done quickly.

Yours truly pulled off the throttle to the engine and then tried to overcome the pilot's "frozen" condition. Seated normally, I could not move the left rudder. That was my cue to "abandon ship" as the Navy says. Opening my canopy and unfastening my seat belt, I started to leave the plane. Then, looking back at the altimeter, I saw that it was still only 3,000 feet. It hit me that I had time to "give it another try." This time, using both feet on the left rudder, I was able to overcome his force on the right rudder, which stopped the airplane's rotation. Centering the elevator came a lot easier. With his full force opposing me, I pulled back on the elevator control and started climbing. As we reached 8,500 feet I could feel his physical input starting to relax. After several minutes he was back in control. He suggested that we land at our Field and have a smoke, which he did.

After finishing his cigarette he suggested that we go back up and finish the last hour of the scheduled training. I said, "Do you think we should in view of what happened up there?" He said, "What are you talking about?" I then told him the whole story. He denied everything and said that I was trying to get him "washed out" of school. With that, we got back in our plane, taxied back to the main parking ramp and called it quits for the day.

Many years later, while instructing in B-47s at Wichita, Kansas, the school psychologist explained to me that a person's state of mind often "shuts down" when faced with a serious threatening situation. He likened this story about the pilot to that of a drunk falling down a flight of stairs. They never remember.

Chapter 4

DEAD STICK LANDING

This story is about the first "close call" I had in a B-24 (see App.p.4). Before going overseas I was assigned to the 459th Bomb Group at Davis Monthan Airfield, Tucson, Arizona.

On one of the training missions we were scheduled for some gunnery practice and a navigational leg. After completion of these, the flight engineer suddenly called out, "Head for home. I'll explain in a minute." I quickly put the plane in a steep bank and headed for Tucson. After flying for only three hours on a scheduled five-hour mission, I couldn't imagine what had gone wrong. After a few minutes of checking, the flight engineer stated that the fuel tanks were apparently empty. There was no fuel in the tanks.

I called Davis Monthan, told them of our problem and that we would arrive over the field in about twenty minutes. As we approached the field there were several B-24s practicing landings. The tower repeatedly told them to depart the landing pattern because of our inbound emergency. My plan was to execute a 270 deg. overhead approach.

We came over the field when number one engine ran out of fuel. On the downwind leg number four engine ran out of fuel. Finally, on the base leg the number two engine quit. Only engine number three was still operating and that throttle was in idle in preparation for landing. As I turned final, another B-24 turned in front of me; he was only 300 feet ahead of me. With three engines out of fuel and engines feathered, the fourth engine was still operating and we had only one hope, if the other aircraft stopped on the runway we would have to move over to the right of him and land next to him on the grass.

As we were about to land, his turbulence threw us into a near vertical bank. I could see the runway landing numbers out of the left window. Putting in full right rudder and ailerons and with the help of the copilot, we managed to right the aircraft. As we moved the plane to the right in the grassy area to avoid the other aircraft, we were still fighting off the other plane's turbulence. Once over the grass, the other plane took off. We then moved left over to the runway and with a few unhappy bounces we settled onto the runway. We turned off at the end of the runway and called the tower to be towed into the parking area.

After parking, maintenance crew checked the fuel tanks with a dipstick and found that number one, two and four fuel tanks were empty and number three tank had only two inches of fuel. A tank malfunction had caused the problem.

Chapter 5

ALL RADIOS OUT

The main purpose of this story is to thank a certain WAC (Women's Air Corp) who trained me in a LINK TRAINER (Simulated Flight Machine). This was a special instrument procedure in case all radios should fail. This procedure was instructed at Ft. Worth Army Airfield.

During my assignment with the 459th Bomb Group at Westover Field, Mass.., we went out for a practice navigational leg. Two hours after takeoff we were about 200 miles over the Atlantic, southeast of Cape Cod, when we discovered that, due to inclement weather, we could not complete any form of a navigational leg. Additionally, we found that all radios were out and cluttered with static. We decided to fly home for one hour to the northwest, followed by one hour to the west. By doing this we hoped to get back over land, namely, Massachusetts. If our calculations were correct we would be over Massachusetts and, still under weather conditions, we would fly on a southwest heading until we ran out of fuel or got a visual sighting of land, we could find our way back to Westover. The southwest heading was to keep us away from the Great Lakes.

After two hours of flight we broke out of the clouds and discovered we were twenty miles south of Albany, New York. Visual flight rules existed in the area so we headed southeast, trying to get under the storm southeast of us and come into Westover, Mass. at minimum altitude from the north. The further east we flew the worse the weather got, pushing us south of our course and bringing us to either New Haven or Hartford, Conn.

The die was now cast. So we flew by dead reckoning due north to Springfield, Mass., heavy rain and all. The fuel was now getting low. Turning to the frequency of Westover, we waited until the ETA (estimated time of arrival) ran out. At that point we could pick up an occasional "station identification signal" for Westover (see Appendix, p.5). A radio range station is very simple and straightforward, providing you have the "A" and "N" signals to help you. In this case we could barely hear the station identifier. Because the static blotted out everything else, most of the radio signals were unreadable. Two station identifiers were broadcast, one every minute, followed by the "A" or "N" signals, whichever radio quadrant we were in. When we were on one of the "legs" of the radio station, both sets of identifiers were broadcast in equal volume. In the bi-signal zone of the "A" quadrant the first identifier remains silent and only the second identifier is broadcast. Likewise, in the "N" quadrant, the first identifier is broadcast while the second is silent. When you hear only one identifier, you only know that you are in the middle of the four quadrants. Crossing the radio station legs was our only clue to help us determine which quadrant we were really in. The bi-signal zone is a sure clue to your position.

After flying past the station for several minutes and receiving mixed signals on quadrant location, we reversed course to the left and flew back several minutes until the letter identifiers got fairly strong. From there we turned to the west. Receiving two constant identifiers we felt that we had crossed the southwest leg of the Westover range. Again, reversing course to the east, we received a constant identifier telling us that we were in mid-quadrant and heading for the station (see Appendix p.5). We had no way of knowing when we were directly over the station, so we flew east the same time we flew west and also received an increase in the identifiers signal as a crosscheck. At this point we took up a heading of 170 degrees and descending hopefully towards Westover.

Five miles from the field, we broke out under clouds and saw the runway straight ahead. It was raining hard. A short distance from the runway we realized we were too fast. We had a tailwind instead of a headwind. With tailwind we used up most of the runway getting stopped. Being very short of fuel and with the bad weather south of the field, we were forced to land downwind.

The WAC (Miss Smith?) that instructed me in this unusual instrument approach and landing procedure was stationed at Tarrant County Airport, now Carswell Air Force Base, Ft. Worth, Texas. It was there that my "initial B-24 training" was accomplished. Miss Smith said that she taught this procedure to very few pilots, mostly because of time limitations and the low priority of the procedure. She also said that she doubted if any pilot would have to resort to using this procedure.

Chapter 6

DOWNBURST

After being hospitalized with diverticulitis for six weeks at Westover Field, my outfit, the 459th Bomb Group, left for overseas so the local personnel office sent me to Boise, Idaho, replacement center.

On the 22nd of May, 1944 we were flying in eastern Idaho when our radio man picked up a message directing a "weather recall" and to return to Boise immediately. From a distance we could see a thunderstorm south of our base. As we approached the base we noticed that the storm was showing strong activity in the immediate area. As we entered the traffic pattern we started experiencing a heavy rate of descent. This was about 1500 feet per minute. Then the bottom dropped out. We had a high rate of descent so I increased power to the maximum. With a further rate of descent, I pulled up the landing hear and went to maximum power. We continued to descend until we reached a point 50 feet above ground level. The aircraft leveled off at this time and started a very slow climb.

Having had a real scare with the storm, we proceeded to land immediately. After we landed and turned off the runway, the storm hit the field with all its fury. The visibility had gone to zero and the aircraft was shaking violently.

It is my belief that this incident at Gowen Field was, in fact, a down burst as we know it today. A downburst is a heavy rainfall within a thunderstorm and associated with a strong downward wind and spreads out over the ground with winds exceeding 60 miles per hour.

Chapter 7

HOT DAY TAKEOFF

Another incident occurred on June 8, 1944 at Mt. Home, Idaho on a hot day. We were scheduled for an afternoon flight takeoff and were second in line for takeoff. While waiting our turn for the runway, we watched the first plane take off. As he rolled down the runway it didn't look like he would make it. From our vantage point we could see that he used all of the runway plus a few hundred feet of dirt and a barbed wire fence at the end. After takeoff he raised his landing gear and then struggled to about 30 feet of altitude.

Back in our plane, we decided that the other pilot must have used improper takeoff procedures. After the tower cleared us to use the runway, we decided we would follow "maximum short field takeoff procedures." We went to the very end of the runway and set maximum power for takeoff. After the tower cleared us for takeoff, we released brakes and started to roll. We lowered the takeoff flaps when we were two thirds of the distance down the runway. This was to minimize air drag until we needed the flaps.

The takeoff roll seemed a bit slower than usual but we felt that we would still make it. Needless to say, we repeated the performance of the first plane. As we went through the fence that the first plane had removed for us, we started to lift off. After pulling up the landing gear, we followed the other aircraft on a very uncertain flight. We just were not climbing and that spelled trouble.

Needless to say, after I reported this takeoff incident to the tower, flying was cancelled for the day. Subsequently, the investigation showed that temperature and pressure altitude computations were incorrect. After this happened, new charts were drawn up.

Chapter 8

TO NORTHERN IRELAND

After completing training at Gowen Field, Idaho, our crew boarded a train for Lincoln, Nebraska, arriving there on June 17, 1944 to pick up our new B-24. Leaving Lincoln, we flew to Syracuse, New York for maintenance work. This consisted of getting new Tokyo tank fuel booster pumps installed. These were to be more spark-proof than the ones removed. After that it was off to Grenier Field, Massachusetts. While we were there we received our overseas equipment. We weren't sure whether we would fly to England by the southern or northern route. The last choice was a flight direct to Northern Ireland. After departing Grenier Field we flew to Goose Bay, Labrador. Our decision was finally made for us when we heard that Greenland and Iceland were "socked in" with weather.

On July 4, 1944 we left Goose Bay for Northern Ireland. Winds were forecast to be a "tailwind all the way." That statement was to come back to haunt us later on. After daily engine run-ups at Goose Bay, we had asked to have our fuel tanks "topped off." The local command refused our request.

About mid-ocean our navigator informed us that we had passed our point of no return (PNR), Goose Bay. We were committed to our destination of Northern Ireland. After passing the PNR, the Navigator took a set of ìsun shotsî and now reported that we had a ìheadwind.î Then, as we flew on, our flight engineer informed us that the fuel in the standpipe gauges showed very little fuel left.

One hour before landfall at the Irish coast, the fuel gauges read ìemptyî and land was not in sight at this point. All crewmembers assumed ditching positions, preparing for the worst, a dunking in the North Atlantic. Also, at this time, we went into ìmaximum cruise fuel consumption.î One hour before our ETA (estimated time of arrival) at the coast, we started a 200-foot per minute rate of descent, and reduced engine rpms to 1,200. We were preparing for a controlled ditching versus bailout.

Arriving over the coast of Ireland, we switched from ditching position to possible bailout position. We then waited for the inevitable. With all crewmembers looking for an Emergency Field, we could not help but notice the very beautiful green landscape of Ireland as it drifted by. It is truly an emerald island!

We were now down to 5,000 feet with clear skies and wondering what reaction we would receive from the Irish Defense Forces. Backing up a moment, we had been informed before we left the states, that to land in Ireland would be against our best interests. That is, unless an emergency existed. Was îno fuelî an emergency? We continued on to Northern Ireland, hoping for a miracle.

Checking weather, we found that the airport had a full overcast. Crossing into Northern Ireland, we established radio contact with Nutts Corner tower. After declaring a fuel emergency, the tower directed us to the published instrument approach. Over the radio station the tower advised us to fly eastbound for three minutes and make a procedure turn, then head back to the field; at minimum altitude the field would be straight ahead. Considering the criticality of our fuel, I wondered, "Where, oh where is my fuel?" We broke out of the clouds at 1,500 feet and, as we passed over the radio station, we sighted the field three miles straight ahead. The fuel held out and we made a safe landing. The miracle had really happened!

We taxied into our parking spot and shut down the engines. On post-flight duties, the engineer crawled out on the top of the wing and dip-sticked the tanks. There was only one inch of fuel in each main tank. That's not enough fuel to make a second landing. Upon further examination of the fuel tanks, it was discovered that the wing tip (Tokyo) tanks, which did not have fuel gauges, were still half full.

Further investigation revealed that the new booster pumps that were installed in Syracuse, N.Y., had only half the fuel flow rate that were in the original pumps. None of us were aware that in accepting a more spark-proof booster pump, that it would take double the normal time to empty the tanks. We were still very fortunate to get to our destination with this lack of information on the new fuel pumps.

Chapter 9

TARGET: BERLIN - NOT TO BE

After arriving in England and having put about ten missions under my belt, the thought kept running through my head, "When will we get a chance to bomb Berlin?" My personal goal was finally realized on January 31, 1945. When that day arrived we were all in the briefing room waiting for the curtain on the mission map to be pulled open. There was a quiet murmur among the crews that this was to be a long mission. When the curtain was pulled aside everyone could see that the feared target was Berlin; there was a distinct change in the way everyone responded. Nobody wanted to talk about it; they knew that this one was the Big One, the granddaddy of all missions. This was the one that we would talk about the rest of our lives, if we lived through it. Mission briefing was extra quiet and all crewmembers were hanging on to every word that was said. This was not a mission in which to forget anything.

After forming up our group, we departed to the east towards Denmark. As we left Denmark our number four engine lost oil pressure and the engine was shut down. Later, as we turned on the bomb run for Berlin, the command pilot looked out his right hand window and observed that we had shut down one engine. He, thereupon, ordered us to return home, bombs and all. He thought that with my presence with an engine out, it would invite German fighters.

Flying back over Germany, the Baltic Sea and Denmark was the longest trip of our lives. There we were, all alone and we had no fighter protection. The third wave of P-51s stayed with the group. All our crewmembers were ordered to stand special watch for any German fighters. They were told that if any fighter came after us we would make a right turn and make a diving high speed run for

Sweden. Could anything be better than the cozy company of those beautiful Swedish girls? Do you think those Krauts would cooperate? No, they were too chicken! Come up, Krauts! Come up! It just wasn't to be but it was a beautiful dream.

Arriving back over the Norwich area, we turned north to jettison our 1,000-pound bombs. The bombardier set the switches for bomb release and opened the bomb bay doors. He hit the "pickle switch" but only one bomb released. Investigation revealed that the bomb on the left rear shackle failed to operate. One of the two releases on the second bomb had failed to actuate. The bomb was hanging at an angle. We then sent the bombardier into the bomb bay, after closing the doors, and had him put the fuse pin back in the nose of the bomb. This was to prevent the nose spinner from arming the bomb. The bombardier took his chest parachute and a large screwdriver with which to attempt to dislodge the bomb.

After getting into position on the narrow catwalk with the bomb bay doors open, he made several attempts to release the bomb shackle but to no avail. We then told him to brace himself and we would do several "dipsy doodles" in an attempt to dislodge the bomb. After accomplishing that, we asked him to try again to dislodge the bomb. It was successful.

Chapter 10

A CASE OF CHICKEN

As we were preparing for our bomber stream departure for Germany, we assembled over a radio station called "Buncher Six." Our group consisted of thirty aircraft. On the second to last circle around the buncher, just as we turned north, another bomb group came into view, heading directly toward us. We thought they would turn to the west where they were supposed to assemble. As our bomber formation came closer together we wondered who would give way to the other (see Appendix p. 6&7).

When it became evident that this was an air born game of "chicken" our leader called out, "Hey, guys, tighten up the formation and maybe we can scare them out of our path." Our planes moved in to a close formation. The other formation came through us like water over spaghetti in a colander. The other airplanes were scattered everywhere! They went through our formation, under our formation and over our formation. They were not able to get their group together and never flew the assigned mission to Germany. A real hair-raising experience.

Chapter 11

HARD CHOICE ON BOMB RUN

On March 17, 1945 our group deployed as individual squadrons to attack targets in support of General Patton's advance into Germany. During our early morning briefing we were told that our route to the target had not been coordinated with the other groups. There was a good chance that our routes might be compromised by these other groups and that, as leaders, we would have to make hard decisions to protect the integrity of our unit and still get the bombs on target. Under no condition were we to circle and make another bomb run.

The mission was uneventful as we crossed France into Germany. While approaching the bomb run it became apparent that another squadron was on a converging path to our target. He was ahead of us on our right and only 200 feet above us. As it turned out, their target was to the left of ours. As we approached our IP (initial point), the two squadrons were about to merge. Our squadron was going to slide underneath the other squadron. At this point their open bomb bay doors nixed that idea so we flew alongside of them and slowed up to let them pass.

You see, we were looking up at hundreds of 500-pound bombs waiting to be released. To get to our target I ordered the other squadron out of our way. They held up on their bomb drop and allowed us to move to the right. We successfully dropped our bombs on target.

Chapter 12

LAST MISSION - ALMOST

My last mission to Germany was on March 31, 1945. We were flying next to the Command pilot, Col. Hubbard. After the usual pre-briefing, with breakfast stuck in there somewhere, we proceeded to our aircraft to start our pre-flight inspection. Takeoff and group form up was normal, as well as the route to the target. By normal, I mean the usual anti-aircraft flack and the usual holes in the plane. Luckily, no one was hurt or impaired.

As we headed back to England we were notified by our "scout," the weather ship, that a lower cloud deck had settled over England, probably less that 10,000 feet. Crossing the English Channel, we could observe the cloud deck up ahead. As we descended in bomber stream formation at an increased rate of descent to get under the cloud deck, it became apparent that we couldn't make it. At this point, we had crossed the English coast. The Command pilot decided to circle back over the English Channel and come in at a lower altitude to be under the cloud deck.

Halfway through this large turn, we noticed that Col. Hubbard in the lead aircraft had taken a strong interest in our aircraft. Was he unhappy with my formation flying or did he see some battle damage on our plane that we were not aware of.

After completing our descending turn, we headed back for our home station. Over the base, we went into the normal break-up procedure, one squadron at a time, followed by peel-offs of individual planes. Landing and parking were uneventful. Upon debriefing we were allowed a beer with a shot of Old Methuselah. This was to loosen our tongues and refresh our memories regarding events on the mission. You can take my word for it, it worked. Being my last mission, it was especially appreciated.

After debriefing was completed, Col. Hubbard made some re-
marks. What followed was a shock to me! He stated that during the
big turn-around over the English Channel, my element leader on
my high right position had set down on top of our aircraft, fitting
so tightly that he could not see daylight between our airplanes. Col.
Hubbard stated that we seemed to meld together for 15 seconds.
No contact was made and we gradually drifted apart. He thought
there would be an imminent mid-air collision and that the Group
was about to lose Lt. Long on their last mission. He hastened to say
that he was afraid to say anything for fear that one of us would
make a movement   in the wrong direction which might cause an
accident.

Chapter 13

BAD WEATHER OVER BASE

In November 1944, while returning from a mission over Germany, our crew was squadron lead, the tower informed us that the weather was bad at our base and that we should go out over the North Sea where the weather was supposed to be clear and bring the squadron down to a low altitude and come in to the base.

Taking the squadron out to the North Sea proved difficult, as the weather was not clear. However, we spotted an opening in the clouds so we echeloned our planes and descended, one at a time. As we approached our field we were at 300 feet and several miles to go. At this time our second crew told us that his radio compass was out and that he was lost in the clouds behind us. Not to be deterred, we told him that we would fire some red flares. In the clouds, our second crew then reported sighting the flares but not our plane.

While this was going on our copilot was calling out magnetic compass heading, altitude air speed readings. When we dropped out of the clouds, the second crew was directly over us!

Chapter 14

GOING HOME - MAYBE

After being released from combat duty, my orders were to proceed to Wales in southern England where I would await transporta-
tion back to the States. Wales is a very unique country, very much
like England, very green and beautiful.

After waiting ten days we were notified that our transport would
be the French flagship, the Ile de France. We knew that it had been
a luxury liner that could cruise at 32 knots and we also knew that it
was too much to expect to still have luxury accommodations. They
had stacked beds to the ceiling of each compartment without too
much space for headroom between bunks.

On our first day out we had a strong wind from the north with
the liner listing to port. It was quite cold, that is, unless you laid on
the deck out of the wind. The trip was supposed to take only three
days but the ship's radar picked up a German submarine so we
changed course and our trip would take five days.

As we approached the Port of New York we were aware that
there might be more subs. At 1:30 A.M., as we were watching the
Statue of Liberty float by, I found that this was the most exciting
part of the voyage.

Chapter 15

CANADIAN DRIVERS

After retiring from the Air Force and finding nothing to do, I put my family in our Chrysler station wagon and decided to drive to Alaska for a vacation and, incidentally, to visit my two brothers who lived in Anchorage and Palmer. We pulled a twenty-foot camping trailer which slept eight. I had my wife and five children plus a dog.

The trip was uneventful until we were westbound on the main Canadian highway. There wasn't much traffic although it was 8:00 a.m. Down the road ahead of us, coming directly at us in our traffic lane, was a large Chrysler sedan. Turning on my headlight to get his attention didn't work so I moved over with my car and trailer on the shoulder of the road. When he reached us head on, I saw a man with both hands on the top of his steering wheel, bleary eyed and possibly intoxicated, driving east into the morning sun. At this point our trailer was almost tipping over.

During the afternoon we had another close call. While traveling west I noticed a highway maintenance worker on a tractor-mower in the grassy area on the left side of the road. As we approached him, he suddenly swerved to the right, crossing the road in front of us. I came to an emergency stop, only 20 feet from his mower.

Chapter 16

AUSTRALIA OR BUST

After returning from Alaska and settling down in Wichita, Kansas, yours truly started climbing the walls. Babysitting the house and kids while my wife worked was not my idea of having fun during retirement.

With all the close calls I've had, you would think I would quit flying. No way! Spotting an ad in the Wichita paper for pilots to ferry small planes world wide, I made a quick trip to the company's office. The first thing they did was to test my resolve by stating, "How about flying a single engine airplane to Australia?" Not to be deterred by their remark, I quickly replied, "How soon do we leave?" I was hired on the spot.

Leaving Wichita, we stopped overnight in Phoenix, Arizona. Then departing from Phoenix and airborne, two things happened with my Beach Bonanza. First, the entrance door popped open and made a terrible racket. Second, fuel started squirting out of the gas cap from the 45-gallon fuel tank that was secured to the co-pilot's seat. Immediately landing and closing the door securely, I took off again with my finger on the air hole in the fuel cap. After arriving in San Francisco, maintenance crews quickly found out that an oil tank cap had been put on the fuel tank instead of a proper fuel tank cap.

The flight departed 17 days later, after completing the mission planning for the next leg of our trip to Honolulu, Hawaii. There were five planes in the formation, four Beach Bonanzas and one Beach Baron. On this day we had an average headwind of 5 miles per hour, which was our maximum for reaching Honolulu with 14 hours of flight time with two hours of fuel left. We had two 90-

gallon fuel tanks, one 60 gallon fuel tank plus a 45 gallon tank fastened down to the copilot's seat.

The morning we left SFO we were assigned an altitude of 10,500 feet. We went in trail, slightly stacked up. After we left the Golden Gate Bridge, we were enveloped in a squall line going to the Farallon Islands. We all had agreed before takeoff that if we should inadvertently enter clouds we should all hold the same air speed, heading and altitude that we had when we entered the clouds. That sounds good on paper but being the last aircraft it was my gut feeling that slowing up a tad would keep me from getting too close to the aircraft ahead of me. I was right. When we broke out of the clouds, the plane ahead of me was only a few feet in front and below my propeller.

But wait, the fun had just started. We had lost our formation leader. We had communication with him but no plane in sight. We told him that we would meet him at Ocean Station November, a navy aircraft carrier located halfway to Hawaii which was a navigational fix.

After eight hours of flight we were at where the ship was supposed to be but there was no ship in sight. Communications with them revealed that they had departed for their gunnery range and further, our leader's radio was fading out.

After ten hours of flight, I advised our leader to turn 45 degrees left and we would hopefully have him rejoin our flight of four. By then it was dark. After 45 minutes of flight, our leader came into view, his navigation lights being our only clue. We successfully contacted him on radio and advised him to contact approach control. I advised him to turn 25 degrees to the left to "home in" on the Honolulu radio. Approach control said that we were not where we should be and advised us to report in at the Delta intersection of radios. This meant turning another 25 degrees left. After reaching Delta we turned right and headed again for Honolulu.

Then it happened. We were flying in a loose diamond formation when a jet suddenly appearing in our midst and focused a spotlight on the Beach Baron whose pilot was, I'll name Phyllis. When the spotlight was thrown on her plane she was reaching down in the cockpit to change the fuel selector. Pandemonium must have broken loose, she was blinded by the spotlight but she needed fuel. She handled it well by changing the fuel selector and at the same time maintained control of the aircraft. The jet took down her aircraft number and left. Phyllis might have said, "That F-102 scared me pea green!"

When we landed in Hawaii, we found that SFO had not forwarded our flight plan to Hawaii and Honolulu had thought we might be a flight of Japanese Zeros coming in from the northeast. Flight time to Hawaii was 14 hours with two hours of fuel remaining.

After two days of rest and sightseeing, we accomplished our mission planning for the route to Majuro in the Marshall Islands. We went by Johnson Island to have a check on our navigational capabilities. We missed it by 100 miles. Something was wrong but we didn't know what. As we headed west and crossed the International Date Line, our leader called me on the radio and said that an Air Force C-124 was behind us and to the left. I am a retired Air Force officer so I got permission from our leader and climbed up 1,000 feet and flew formation on the C-124's right wing and then called him on the distress frequency and asked him for a navigational fix. Faces started appearing in the C-124 windows. They couldn't believe that a Beach Bonanza could possible be out there in the middle of the Pacific Ocean. After much cajoling and explaining, they gave me a fix. Returning down to my formation I advised our leader that we were off course by 200 miles. We then made a left correction for this error. Later we made a 20-degree correction to the left to reach the island of Majuro. Without the C-124 we would have wound up in a very wet spot in the Pacific.

After a three day delay in Majuro, waiting for a cyclone to move out of New Caledonia, we fueled up with what gas they had. They did not have enough fuel with reserve to reach New Caledonia. Now the fun starts. The leader took off with no operable radio receiver. We found this out after takeoff. He had an operative transmitter. I moved up and flew formation on his right wing. After six hours of flight, I noticed an unusual shaped island on my National Geographic map. We were 200 miles off course to the right. If we had stayed on this course we would have run out of fuel in the middle of the Coral Sea, namely Sand Island which, I believe, was only a sand bar. We needed to turn 45 degrees to the left to reach Port Vila, what with minimum fuel and completely off course, we couldn't possibly reach New Caledonia.

Getting the leader's attention was difficult so I moved up in front of him to encourage him to turn left.

He said, "Do you want me to turn left?"

I wagged my wings indicating "yes."

He said, "Ten degrees?" I didn't move.

He said, "Twenty degrees?" I didn't move.

With desperation he said, "Thirty degrees?"

I wagged my wings meaning "yes."

We were finally on course for Port Vila. But wait, we were flying on the east side of a row of thunderstorms. This was no place for good aviators. At this point, our leader suddenly made a hard right turn and disappeared between two thunderheads. As deputy leader, I refused to follow and so advised the formation. They stayed with me. After the thunder storms, Phyllis elected to fly out in the direction the leader had gone but could not find him. That evening we received a message from our leader. He had landed in New Caledonia and ran out of fuel just as he turned off the runway. He had to be towed into a parking area.

The next two days took us to New Caledonia and Australia.

Quizzing the leader about his errant compass headings, he said he "swung his compass" at his airport 30 miles east of Wichita, by flying down farmers' fences but using the prevailing winds from Wichita Airport, which was west of Wichita, a 40 mile discrepancy. The winds can be very different at that distance. A compass rose is a fixed sign painted on a taxi ramp. As an airplane lines up on the compass nose bearing, the pilot reads and makes a notation of the compass reading error. This gives a pilot the deviation factor to add or subtract from his aircraft heading. We returned from Australia via commercial air.

Chapter 17

FOGGY LANDING

After returning from Australia, I joined the Idaho Air National Guard at Gowen Field which was a fighter interception group flying P-51 Mustang fighters. This was the same field that I had earlier been stationed at flying the B-24s. During the summer of 1950 the squadron moved to Walla Walla, Washington for maneuvers. Thereafter, we planned a return trip to Boise, Idaho with a full complement of aircraft. We would arrive there in a tight squadron formation and land.

The first thing on the agenda was to climb over the mountain range en route to Boise. Approaching the range, the leader picked out a low spot (mountain saddle) and hoped that we could fly over it without circling to attain a higher altitude. Yours truly was flying in the lowest part of the formation. The closer we got, the less sure it seemed that we would make it. In fact, with only miles to go, the lead plane put down his flaps to gain more altitude and waited for the pine ridge to come us. To make a long story short, we cleared the trees. Just barely!

After crossing the mountains, our leader contacted the tower and was told that fog had moved in and that landings at Gowen Field would be touchy at best. From our altitude we could see that the fog had moved into the Boise area and might interfere with our landings. We went beyond the field and executed a 180-degree turn. Before we reached the field the fog had covered half of the airport runway. We would have to land on the first half of the runway and coast into the fog. We had no choice. Nearby Mt. Home airbase was closed.

As we came over the field, each plane peeled off in its proper order and landed. However, by the time the last plane was landing only 2,000 feet was left to land on. Yours truly made this short landing and coasted into the fog. Visibility was better than expected in the fog and I coasted up the runway to the correct taxiway and turned off into the parking lot. End of another thriller!

Chapter 18

BENT WING

While flying with the National Guard, we were accustomed to the rip roaring powerful P-51 with its Rolls Royce engine. However, on the other days, we flew the two-place T-6 to accomplish instrument training. In this setup, the pilot in the back seat would pull back the instrument hood so that he wouldn't have any outside references for flight purposes.

One day a buddy asked me to fly "safety pilot" in the front seat while he practiced "instruments" in the rear seat. After preflight inspection was complete, we crawled into our respective seats and cranked up the engine. As we were about ready to taxi out, my friend in the back seat asked me if I saw a wrinkle on the left wing. After eyeballing the left wing and comparing it to the right wing, I stated that no wrinkle was observed on the left wing. Of course, my position was forward of the other pilot. He had a low, angular view. He decided we should proceed as planned. It was his plane.

After being cleared for takeoff, we climbed out to the instrument training area. After leveling off, my friend put the plane in a straight and level flight and tried to adjust the trim tabs accordingly. After several tries, he got frustrated and asked me to trim up the plane from my position. After several attempts, I gave up and informed him that the plane could not be trimmed. He then decided that we should return to Gowen Field and forget the exercise. We landed uneventfully, parked the plane and went home.

The next day the Air Guard Commander called me up where I worked to ask me why and how I had wrecked one of his airplanes. I countered by telling him that when we went up to practice instrument flying, the plane wouldn't trim up so we returned to base and parked the aircraft. His next action was to call up the pilots that flew the airplane the day before we did. He hit the jackpot!

The story that these pilots related was absolutely unbelievable. It seems that while they were up flying the plane, the pilot in the back seat wanted to take a "breather" from instrument flying and told the safety pilot to take over the controls of the plane. Thereupon, the plane went into a gradual dive, angling over into a steep dive. With the ground coming up fast, each pilot thought the other pilot was carrying the dive-bombing too far. Both pilots then grabbed the elevator controls and pulled back with all their strength, narrowly missing a crash in the desert. They decided to call it a day and landed and parked the plane.

Another pilot intending to fly the next day noticed on his preflight that the left wing was bent upwards and could see the wrinkle in the wing. The wing was bent upwards at a point halfway out on the wing. This turned out to be a double whammy. The original pilot had accidentally bent the wing when they pulled out of the dive. The squadron commander inspected the aircraft and ordered it "grounded." It would never fly again.

Chapter 19

CHECK RIDE

On November 12, 1950, while flying AT-6s with the Guard, I had to take time out for my annual instrument check ride. The check ride was very comprehensive and I "aced" all parameters. However, the check pilot decided that he would give me one more test, even though the check ride was complete.

As with all check rides "under the hood" I caged the flight indicator and the directional gyro and closed my eyes. The pilot said he would give me another "unusual positions" check. He started by going through various maneuvers in an attempt to disorient me. At the appointed time he said that the plane was all mine. Opening my eyes, I determined that the wings were level, the rate of climb high and the airspeed was dropping off fast. The only proper action was to push the nose forward quickly to recover airspeed before a stall occurred.

As I pushed the nose forward there was a loud commotion in the front seat. After several seconds, the pilot said, "I've got it" meaning that he was taking control of the airplane. After we landed and were walking back to the hanger, the check pilot said that he would consider this flight a training ride. In other words, he was technically failing me on my check ride without telling me why. Several days later, the operations officer sent me up to retake my flight test (a C-47). As it turned out, I passed.

Several months later, I heard through other pilots that when I pushed the nose down to recover airspeed, we were actually upside down in the middle of a loop. The noise in the front cockpit was unfortunately the check pilot crumpled up in the top of the plastic canopy. It seems that when I pushed the nose of the plane forward, his safety belt was not buckled up and he wound up, you know where, in the canopy of the aircraft.

Chapter 20

HIT BY LIGHTNING

On April 1, 1951 our fighter group was recalled to active duty and moved to Moody AFB in Valdosta, Georgia. During the month of June, having little else to do, I wandered down to the flight line to watch the planes fly off the airport. Because of the sweltering heat and humidity, I took cover under a shade shack that protected the fire trucks from the elements. While standing there, we experienced a heavy cloudburst. The shelter worked fine with its dry soil to stand on. The rain came down so hard that it flooded the area we were standing on. There were five of us under the shelter standing in the rainwater and not having enough sense to crawl up on the fire truck, we were suddenly vulnerable and didn't know it.

Then it happened. A bolt of lightning hit the top corner of the nearby airplane hanger about 100 feet away. As it traveled to the bottom of the hanger, the lightning spread out on the wet ground like a spider's web in all directions and raced across the ground to our shelter. As the lightning hit us, we suddenly crumpled to the ground in a fetal position. (See Appendix 9) Seconds later we all stood up and congratulated ourselves on still being alive to talk about it. That was some jolt!

Two minutes later it happened again. Another lightning bolt hit the same spot on the hanger, raced across the wet ground and hit us. As before, we reverted to the fetal position and seconds later we were again congratulating ourselves on missing the bolt. Who says that lightning doesn't strike twice in the same place? This was our answer.

Chapter 21

APPARENT LOSS OF AIRSPEED

While flying B-29s out of MacDill AFB, Florida with the 305th Bomb Wing, I had an opportunity to go on a flight to Fairchild AFB at Spokane, Washington. After a two-day stay, we filed a clearance to MacDill. The weather there was rather sickly looking but good enough for a "white card" holder, which I carried. The preflight, engine run-up and taxi-out were all normal. While waiting for takeoff instructions, the tower advised us that the local weather had dropped to instrument weather, a ìgreen cardî, which I did not possess.

After much radio communication with the tower and our operations at MacDill, it was decided that the multi-rated pilot-navigator would be in command of flying the aircraft and I would move to the copilot's seat for takeoff. When all seat changes were made, we asked for a takeoff clearance. The tower reported the takeoff conditions as low clouds and foggy with temperature close to freezing. We were cleared for takeoff.

Advancing power, we quickly reached takeoff speed, retracted the landing gear and as we were passing through 300 feet, the airspeed suddenly dropped to zero. At this point the pilot quickly pushed down on the elevator to regain airspeed and I, as quickly, told him to "pull up." As I told him to "pull up" I turned on the pitot heat and in a few seconds the airspeed returned to normal. The pitot tube had suddenly frozen with moisture and had blocked air flow to the airspeed indicator.

Arriving hours later at MacDill, they reported the field was closed due to fog. We thereupon diverted our flight to Miami. After a three-hour delay at Miami, we took off again for MacDill. The weather was reported to be clear within one hour. Arriving over

MacDill, we found that the field was still fogged in. Always the opportunist, I saw that one runway on the east side of the field was clear of fog. Knowing that they wouldn't permit us to land because of Air Force flying regulations, I decided to land on this runway. Turning off the end of the runway, it became obvious that we couldn't see to taxi. We called for a tug to tow us to the hanger. All's well that ends well!

Chapter 22

MARCH AIR FORCE BASE

March AFB was one of the nicest and most beautiful bases in the Strategic Air Command. It was here that one of my most unusual flying experiences took place. Some time during the summer of 1952 a fighter aircraft was practicing simulated attacks on a B-29. This was a head-on attack. Something went wrong: the fighter failed to pull up and created a mid-air collision with the B-29. Many of the crew bailed out or were thrown out of the plane by the force of the impact. There were casualties.

Two months after this accident four airmen from this crew were assigned to fly with me. It was my intention to fly the smoothest and most trouble-free ride possible just for these four men, knowing that they might still be shaken up by the accident. Our first flight was yours truly leading a six ship formation.

Shortly after leveling off at our assigned altitude, all four engines suddenly quit. There had been no warning that anything was going wrong. All four engines had quit simultaneously. No sooner had the engines quit than the flight engineer decided that it was time to bail out. He opened the escape hatch, unfastened his seat belt and was about to leave the aircraft. As he was about to do so, I reached back, grabbed him by his shoulder harness and shouted at him to restart the engines using the "emergency override switches" on his engineer's panel. He obeyed the command and got all four engines running again. We thereupon rejoined the formation and proceeded on our mission as planned.

After landing, the four airmen went into the Operations Office and resigned from flying duty. That was a natural reaction that I could accept.

Chapter 23

PILOT ERROR

During July, 1952, our crew took off in our B-29 for a six hour mission from March AFB, California. Turning off to the northwest and flying over Riverside, CA, then turning northeast to over-fly Lake Arrowhead, we proceeded on course. Over San Bernardino, CA and climbing, it appeared from a distance that we would successfully over fly Lake Arrowhead. So we churned on towards the mountain until my radar operator called me on interphone to warn me that we would not clear the mountains we were approaching. When I received this warning, it appeared that I had committed myself and could not turn left or right without endangering our plane in a steep turn so, with maximum power, we headed for the tree line around Lake Arrowhead. (See Appendix 12)

As we went over the tree line, we were 200 feet above the trees with Lake Arrowhead very visible. By this time, I had broken out in a profuse sweat and wishing I had not been so foolish in estimating the amount of altitude I should have had. If the crew had bailed out of the plane, I would not have blamed them.

Chapter 24

B-36 PROBLEMS

In April of 1953 I was stationed at Carswell AFB going through B-36 transition. With my low flight time, I was assigned as the first officer, normally called copilot. These airplanes had six reciprocating engines and four J-47 turbojet engines (appendix p.10). We used all ten engines to climb to altitude. This story is about a nighttime mission that would normally be of a fifteen-hour duration. We started with a navigation leg, then heading for the gunnery range in the Gulf of Mexico, off the coast of Texas. By the time we reached the gunnery range, number three engine had been shut down due to a malfunction. Not to worry. It was standard operating procedure to lose one or two engines on a typical flight and still continue flying.

Going southeast on the gunnery range, the second engine quit. At this time, we notified Houston radio center that we had two engines inoperative but that we would continue our mission. They were excited in the center and wanted to know if we wanted to land at San Antonio, Texas. We declined their offer. Their problem was that, due to security reasons, we could not tell them that we were a B-36. As we proceeded northwest on the range, the third engine quit. We again notified Houston center of our situation. This was the only time I ever heard radio control get excited. He could hardly talk. Again, we turned down their offer to land in San Antonio. The pilot instructor felt that we should still head back for Ft. Worth, Texas.

En route to Ft. Worth, the two reciprocating engines of the left wing were inoperative and the fourth engine quit. Paradoxically, we only had one fuel pump on the left wing and with the jet engines started and running at full power, this put us in a predicament, if that fuel pump quit, we would have been forced to bail out. No power on the left wing.

Approaching Carswell AFB, we started the last engine that quit, as it was only the air induction system that had failed at altitude. But before we could land, another reciprocating engine caught fire. We discharged the only two engine fire extinguishers we had but that failed to put out the fire. Now we had to land. The tower had alerted all fire departments so we had plenty of fire protection on landing. We landed ok and left the plane on the runway.

Chapter 25

ASLEEP AT THE CONTROLS

From Ft. Worth, Texas our B-36 crew was transferred to Limestone AFB, Limestone, Maine. This was a cruel trick to pull on someone who had joined the group because it was destined for El Paso, Texas. But strange things have come out of Washington, D.C.

Flying out of Limestone required us to fly one of two missions, either a 25-hour mission or a 15-hour mission. On the 25-hour mission there was one serious drawback, the communal toilet would fill to overflowing. What to do? Stop at MacDonald's? On one 25-hour mission, the automatic pilot that was supposed to fly the plane became inoperative and we had only two pilots with four pilots needed. This meant taking turns sleeping in four-hour segments.

Yours truly had been flying the plane for four hours when the aircraft commander woke up from his four hour nap. When he did he sat up in his seat, had a cigarette and then told me I could now take my nap. At some point in my rest period, I was awakened by one of the two flight engineers. The first thing I heard was "Who is flying the airplane?" My answer, "The pilot." Looking over at the pilot's seat, we saw that he was sound asleep. The flight engineer then told me that both flight engineers had fallen asleep and they could not wake up any of the crew. It was a startling fact: the entire crew of fifteen airmen was sound asleep. There was no one flying or operating the plane!

At this time, Cleveland center radio was calling us. Putting on an act of innocence, I told Cleveland center that we had had radio problems and couldn't reach them. They replied on their radar screen that we were flying a large triangle, which is a sign of distress. This was a very stable airplane to say the least. They replied that they observed we had flown three very large circles. Obviously, a west wind at our altitude had misshaped the circle, which they had mistaken for distress. Our plane had been in a five-degree turn for quite some time. Assuring Cleveland center that we now had our radios fixed, I informed them that we would like to pick up our flight plan and continue our mission. It was my feeling that any pilot that fell asleep at the controls should be court-martialed. Nothing was ever reported on this incident and, likewise, the guilty pilot went unpunished.

Chapter 26

A COWARD ON THE CREW

Late one evening of August of 1953, our B-36 was flying a night mission within the states comprising the northeast region, which includes Maine. This flight was out of Limestrone AFB. After five hours of flight, a gunner called from the aft section of the plane and reported that the number four engine was on fire. The pilot ordered the flight engineer to put out the fire with the self-contained fire bottle. When that failed to put out the fire, a second bottle was discharged but to no avail. With that we headed back for Limestone, reporting to the tower that we would be there in fifteen minutes, fire and all.

As soon as we landed, the pilot hit the emergency switch and the plane came to a fast stop. Once the aircraft commander hit the switch, and the plane stopped, the aircraft commander was out of his seat and headed for the exit. Before I could get out of my seat, the aircraft commander was seen running like a deer across the runway to the open field area. And I was dumb enough to believe that the captain should be the last man out. As it was, I waited for the flight engineers to go out ahead of me. When I went through the radio compartment, there was a rustling sound that seemed out of place. Asking if there was anyone in the compartment, I got a reply from the radioman. He said he was trying to find his jacket. After finding his "precious" jacket, we both exited the plane just as the fire trucks arrived. All crewmembers were accounted for, so we watched the fire trucks extinguish the fire. The plane was not destroyed or burned up, only damaged.

Chapter 27

ALL ENGINES SILENT

On another flight in the ten-engine B-36 intercontinental bomber, we were faced with an unusual malfunction. Cruising at 15,000 feet, we entered a storm area west of the Mississippi River and were inundated with heavy rain showers. This was a normal happening on a mid-altitude flight. However, our comfort level was soon shattered.

All electrical power had suddenly stopped. The six reciprocating engines continued to operate as if nothing had happened. Being a daytime mission, we at least had plenty of light. When the power outage happened, the two pilots turned to the flight engineers to find out what happened. The flight engineers were unable to explain. All instruments were dead. The navigator, radar, and radio operator came up from their compartments to find out why they had lost all power. We could not give them any explanation.

After thirty minutes of flight we broke out into daylight, the storm now being behind us. A few minutes later, the electrical power came back on. All equipment in the aircraft started operating again. We were all scratched our heads, we just couldn't figure this one out. After landing, we talked to the engineering officer who, in turn, called the factory at Carswell AFB to obtain some kind of reason for our engine operation. They called back about fifteen minutes later. They stated that the rain water, which is pure at 15,000 feet, had gotten into the alternators and formed an insulating layer between the metallic brushes and the magnets and had effectively stopped the alternator from generating electricity. After leaving the rainstorm, the air at that altitude had dried up the rainwater in the alternator and allowed it to properly operate.

Chapter 28

TWENTY THOUSAND POUNDS TOO HEAVY

After leaving Limestone AFB and being assigned to Biggs AFB at El Paso, Texas, I was ecstatic about flying the B-47 Stratojet bomber. Taking further training locally required me to fly with a professional instructor pilot (IP). On this particular date it was my intent to "fly by books," to do nothing wrong that would reflect on my ability to fly the plane. Everything went along fine until we rolled down the runway.

This was a hot day in Texas so it wasn't surprising that we seemed to use up a lot of runway. As we crossed the last 2,000 feet of runway, it became apparent that something had gone wrong. It was too late to stop so I pulled the aircraft off the runway and raised the landing gear. We narrowly missed having to jettison the heavy wing tanks. We were airborne but not climbing. The sagebrush stayed right below us, and the terrain was slowly rising. As we struggled to gain a few feet of altitude and also make a left turn at ten knots above stalling speed, the IP discovered our problem. He noted that we had 20,000 pounds of fuel in the auxiliary fuel tank, which was supposed to be empty. Somehow I had missed that on my preflight. (See P.10)

Chapter 29

NEAR MISS

During the summer of 1958, while flying the B-47 on a training mission over Texas, we had an unusual happening that can never be forgotten. We were flying west after crossing over Louisiana and were about 30 minutes into Texas when we faced a possible mid-air collision with an unknown aircraft. My copilot called on interphone and said there was a plane in the distance, level with the horizon and at our ten o'clock position. He thought it might be a B-47. After watching the plane for several seconds, we realized it was coming at us extremely fast.

It was difficult to tell whether it would go below us, above us, or hit us. We didn't have long to wait to find out. In seconds it was upon us. It was a B-58, which travels faster than twice the speed of sound. He passed just under the nose of our aircraft (appendix p.11). As it went under us I got a look at the pilot's face. The B-58 missed us by twenty feet. It took six seconds for it to disappear on the horizon to our right side.

Then suddenly we were hit on our left side, as if a Mack truck had hit us. The navigator, who sits down in the nose of the aircraft, which we call "the hole," suddenly called the pilots and added, "What's going on up there?" It took us about a minute to figure out that we had been hit by the B-58's supersonic shock wave. What was so surprising to us was that the shock wave was so far behind it. We had never heard of a shock wave at 35,000 feet while in another plane.

Checking with Air Traffic Control didn't solve anything. I reported a "near miss" but they said they had no aircraft on their radar scope. After landing and reporting the "near miss" we found out that Air Traffic Control had set up a high-speed test track in Texas, running south to north over the states of Texas, Oklahoma and Kansas for purposes of the B-58's high-speed test program.

What was so amazing to us was that ATC knew about it and let us fly blindly into that area. Even though this was considered a "top secret" corridor, there was no excuse for not changing our altitude. A mid-air collision would have spread our wreckage over at least three states.

Chapter 30

HOME FOR AN ANNIVERSARY

During May, 1958, while stationed at Biggs AFB, El Paso, Texas, I was an unwilling participant in my navigator's promise to be home for his and his wife's wedding anniversary.

We were flying our B-47 on a flight encompassing simulated bomb drops on a radar site and a two-hour navigational leg. As we finished our mission over Lubbock, Texas, we noticed a lot of blowing dust in the area. Sensing that the weather had changed for the worse, we headed for El Paso. While over Guadeloupe Peak, we called our control room to get the latest weather report at our destination. They said the weather had suddenly changed and that we should land as soon as possible. Being only ninety miles from Biggs, we started descending for a landing.

While flying inbound, we were advised that the winds had shifted remarkably and that we might not be able to land. Entering the traffic pattern, we were bounced severely by wind gusts. The wind was kicking up large dust clouds and appeared to be coming from the northwest. Turning for the landing, we realized that it might not be possible to land. The cross wind at 45 knots required both pilots to view the runway out the side of the canopy instead of the front windshield. Using maximum crosswind landings techniques, we gingerly let the landing gear touch down on the runway. When we did that, we were rewarded with a violent yank of the airplane. The wind was a direct cross wind.

Realizing that it was impossible to make a safe landing, I applied full power and informed the tower that we couldn't land and that we were headed for the only safe airport in the southwest U.S. that was within our wind and fuel range, Tucson, Arizona. Arriving at Davis Monthan airfield, we declared an emergency for low fuel. Other aircraft were ordered out of the traffic pattern and we landed without further incident. There was only enough fuel in the bottom of the tank for one more trip around the traffic pattern.

After parking the plane and shutting down the engines, the navigator thanked me for not landing at Biggs AFB, El Paso, Texas, even though it was his wedding anniversary and he and his wife had planned an appropriate party. Better safe than sorry!

Chapter 31

NORTHERN LIGHTS

Throughout my life the northern lights have always intrigued me, how they glow and dance in the northern sky and how they form curtains of light that are constantly changing throughout the night. All the colors of the rainbow are illuminated. Not once in my life did I ever think they could cause problems while flying.

One mission we participated in from El Paso, Texas was a "two aerial refuelings sortie" from El Paso to Alaska and return. We refueled over Spokane, Washington with the second aerial refueling over the Yukon River in Alaska. As we approached our letdown point over the Yukon to connect with our KC-97 tankers, we could see the most unusual interpretation of northern lights. The apparent horizon was tilted at forty-five degree angle to the right, while we were actually flying straight and level.

When we completed our descent to refueling altitude, the co-pilot started warning me that if I didn't bring the plane back level, we would surely spin in and crash. It was obvious that he had a case of vertigo. Calming his fears wasn't exactly easy. It is hard to want to believe that your flight instruments can be relied upon but if you are not getting warning flags on your instruments, you must believe them. The copilot finally settled down.

A second long range mission found us flying westerly out of Goose Bay, Labrador. In this case it appeared that we were flying at a forty-five degree angle of bank to the left. (This, again, was due to the northern lights.) Then suddenly the copilot started screaming on the interphone that we should bail out because, again, the apparent high angle of bank gave him a false sense of vertigo. He believed positively that we were surely going to crash. I gave him a

stunning "Shut up! We are in level flight and are not going to crash."
This order had to be repeated the third time before he came to his
senses. It was very difficult for him to believe his instruments. He
had it bad.

This, then, was the second time this copilot had experienced a
bad case of vertigo. Besides, it was a minus thirty degrees Fahren-
heit outside and, had he bailed out, he would probably have frozen
to death before he could be rescued on the ground.

Chapter 32

HIGH SPEED TEST FLIGHT

While flying B-47s at Biggs AFB, El Paso, Texas, one of my duties was to test fly different B-47s. On this particular day the maintenance section had accomplished some rigging work on the flight control system and needed a test flight to confirm that all cables operated properly.

The test flight required the plane to be flown at the maximum speed. Checking the performance charts, I determined that the plane had to be flown at 16,500 feet at an altimeter setting of 29.92 inches of mercury. Getting clearance from Air Traffic Control to give us a block altitude from El Paso to Tucson, Arizona of two thousand feet, we advanced all six throttles to full power and headed west.

As we reached an indicated airspeed of 440 knots, we tested the ailerons by turning the controls both left and right. The plane would not turn in either direction. We had reached what is called "cross-over speed." This was a normal and correct reaction at that airspeed. Then, increasing power and reaching 465 knots (at this speed the mach indicator was at max) we tested the ailerons for a left turn and the aircraft turned to the right. This was the correct reaction. Then, returning the controls to center, we turned the ailerons for a right turn and the aircraft made a left turn. This was also a correct response. Then, centering the ailerons again, we reduced power to normal cruise.

In the foregoing test, the reason the plane turned in the opposite direction from the controls was that, at that speed, "above 440 knots" the wings warp (changes shape) and forces the wings into a position of opposite control.

There are two things dangerous about this flight test. First, if we had to avoid a midair collision, it would be very difficult to make a change of course with the controls in the reverse mode. If a change of altitude was necessary it would put us in a "stall" category with uncertain results. Secondly, if I had let my fighter pilot instincts take over by executing a "chandelle" (pulling up hard and attempting a half loop) the plane would have been put in a high-speed stall, again, with disastrous results.

Chapter 33

LOSS OF NOSE CONES

There was always a mystery when flying the B-47 as to why, where and how, or when we were losing the nose cones off the front of the engine. Mechanics tighten the clamps that held the nose cone in place and Boeing kept coming up with new "fixes" to solve the problem but to no avail.

One day, while returning from a six-hour mission, I was getting ready to start our descent into El Paso. After making sure that there was no conflicting air traffic, I reached over to the six throttles and, at the same time, observing the engines on the left side of the plane. As the throttles reached "idle" position I noticed the number three engine nose cone suddenly shoot forward about five feet and then turn upwards and head backwards like a speeding bullet. (Appendix p.11) Evidently, the back pressure from the engine forced the nose cone loose from its mountings. The mystery remains: how did the nose cone clamps come loose and allow the nose cone to leave the engine? This question was never resolved as the group transferred to Plattsburgh, New York, and the subject never came up again.

Chapter 34

A PISTOL IN MY FACE

During 1957 I was flying a regular practice mission at Biggs AFB. As was our custom, we always kept a pistol with us. Protecting the classified bombing equipment was an absolute must. After completing a late night mission at 2:00 A.M., I found that the gun safe was locked up for the night. Not to worry, I'll take it home with me and hide it so the kids won't find it.

Arriving home, I proceeded to put my pistol in the dresser drawer under all my clean clothes. The kids are not supposed to be in my room anyway. But just to be safe, I put my clip of ammunition on the top shelf of my closet behind everything else up there. Now I can sleep peacefully.

The next morning I awoke with my ten-year old son straddling my chest, facing me with the Colt 45 pistol pointing at my head. For a moment I shuddered. Had he also found the bullets? Talking to him quietly and at the same time moving the pistol away from my face, I asked him to get off my chest. Then I secured the weapon and checked it out. He had not found the ammo. A lesson learned the hard way!

Chapter 35

HARD LANDING

On another late night mission, when we were over flying Chicago, I found that I was starting to fall asleep. Asleep at the wheel! My head was literally falling off of my shoulders. Putting some music on the radio did not help. I asked the copilot to fly the airplane while I caught some zzz's. He replied that he was having the same problem. Getting our notes together, we realized that the Flight Surgeon had given both of us some medication for stomach upset. The problem was that he should have grounded us from flying for taking this medication.

Now, it was obvious that we had a serious problem on our hands: how to stay awake and get the plane back to our base at Plattsburgh safely. We changed course and decided to head directly to Albany, New York and then to our base. By continually talking to each other about everything from soup to nuts, we managed to stay awake even though our heads were trying hard to nod.

Over Albany, we called for landing instructions at Plattsburgh and received a #1 for landing clearance. Chopping power over Albany, we made a beautiful straight-in descent. Coming in over the end of the runway, we rotated the plane for a landing. But wait! We didn't land. Where was the runway? Only too soon did we find out. Due to our depth perception error from the medication, we had landed twenty feet too high and stalled out. When we hit the runway, my chest and head hit my knees hard. Then came the bounce. At this point I managed to bring up the throttles to save the second bounce. As the power came up the plane settled on the runway. A very ignoble landing!

Chapter 36

NEAR DISASTER

While stationed at Plattsburgh AFB and flying the B-47 bomber, I experienced a horizontal wind-shear which forces the plane towards the ground. On this particular day, we had flown a six-hour exercise that included an aerial refueling, celestial navigation and radar bomb drops on the simulated Montreal, Canada radar bomb site. Completing that, we headed home in a well-spaced bomber stream.

Calling in to the Plattsburgh tower, I was instructed to make a Ground Control Approach (GCA) for a landing to the south. Letting down from altitude was an easy affair under cloudless skies around 2:00 A.M. The field was in sight with 20 mph head wind at landing. A no-sweat approach.

Following GCA instructions, the approach appeared normal until we passed through 300 feet. At that point, I sensed a sinking sensation but the rate of descent showed a normal rate of 600 feet per minute. Checking with GCA they reported that I was on the correct glide slope. I immediately went to a 100% power, dropped the approach chute and waited for the inevitable.

A 2,000 foot area was under reconstruction. The construction crews had dug out the entire 2,000 by 300 foot area down to a depth of three feet. This would be an imposing hazard if one landed too short. As we crossed the riverbed a quarter mile from the runway, the sinking feeling accelerated. With maximum power, I could only hope and pray that we could reach the runway. As we crossed the approach end of the runway, I pulled back on the elevator so that my speed was down to five miles per hour above aircraft stalling speed, saving a few feet.

As we crossed the construction area, I looked out to my left and it appeared that the front landing hear had crossed the lip of the ditch. As the rear landing gear hangs lower than the front gear, I could only hope that the rear gear would clear the landing area lip also. Luckily for us, both landing gears had touched down on concrete. Then, deploying the brake chute, we slowed to a stop at the end of the runway and turned off. While jettisoning the brake chute we noticed a fireball two miles away at the construction site. Out of this fireball came the next B-47 a few feet above the runway.

That plane had landed in the construction area with the forward landing gear bouncing out of the construction area and the aft gear touching down in the last few feet, hitting the concrete abutment. This, in turn, ripped the landing gear from the plane's structure and, at the same time, tore open the aft main fuel tank, spilling 10,000 pounds of fuel into the area. This fuel exploded into a huge fireball as the lighted smoke pots around the construction area ignited the fuel.

The pilot thought that it was too hard of a landing and applied full power to "go around" and try for another landing. But then the copilot reported to the pilot, "I think we are on fire!" The pilot then chopped all power and slid or skidded off the right side of the runway and came to rest in the dirt area. After removing the canopy, all three crew members ran down the wing to safety as the outrigger gear had collapsed and the wingtip was touching the ground. The fire truck arrived and put out the residual fire where the aft main tank had been.

Investigation revealed that there had been a strong tailwind on the descent from altitude, which persisted down to 300 feet. This required the six jet engines to be at minimum power to stay on the GCA glide slope. Passing through 300 feet, the wind direction had switched 180 degrees to the south and caught the engines at low power. Full power was the only remedy to reach the runway but the jet engines were notoriously slow when it came to accelerating from idle power to full power. The horizontal wind shear was almost too much for the engines to handle and was almost the undoing of the crew.

Chapter 37

AN ICY RUNWAY

In January of 1959 when the 97th Bomb Wing closed down, our crew was transferred to Plattsburgh AFB, New York. The cooler air was a welcome relief from the heat at El Paso, Texas. While stationed there, we had several incidents that warrant writing about.

On one mission, while making simulated bomb drops on the Montreal, Canada, bomb site, we were informed that our runway was covered with ice at Plattsburgh and we were ordered to divert to Homestead AFB, Miami. A nice two-day vacation followed.

On still another mission, we were arriving back from a training flight, when we were notified that the runway was covered with ice from a freezing rain and that no other field within fuel range was above weather minimum. No time for the faint of heart. We had to land! Remembering back to my teen years and how I learned to drive a car on icy roads, I decided I could safely land the aircraft. One thing in our favor was that there was a direct headwind. Landing the plane in the first 1,000 feet was successful and then, riding the crown, or high point, of the runway, I gingerly kept the plane on course to the end of the runway and turned off at the taxiway. Realizing that I couldn't taxi anymore because of the slippery conditions, I called for a tow.

Chapter 38

INVISIBLE AERIAL REFUELING (KC-135)

While instructing B-47 students at McConnell AFB, Wichita, Kansas, I always instructed my students to treat the planes like a lady. That is, when turns or changes in altitude were made, be gentle; don't try to overpower the aircraft. As a demonstration, I had a student trim up the airplane controls so that it would fly by itself. I would then tell him, while holding the controls, to "think" that he wanted to increase his altitude but not move the controls. As a result, the plane started climbing, then I had him implement the reverse process, to "think" that they wanted to level off and then, descend again. It worked. The students were amazed at how little effort it took.

While returning from Spain after a tour on "alert" status, I was scheduled to meet a KC-97 tanker well out over the ocean after departing the British Isles. Finding the tanker and getting "hooked up" to start the aerial refueling was as easy as pie.

The tanker had transferred about half a load of our scheduled amount of fuel when we suddenly entered a thick layer of clouds. The clouds were so thick that we could not see the tanker or the refueling boom. (Appendix p.12) Normally, the regulations call for a "break-awayî maneuver to separate the planes and avoid the possibility of a mid-air collision.

I was so comfortable and confident on the refueling probe that I chose not to call for a 'break-away." Instead, the boom operator continued to give instructions, "up two or down two or same amount for either left or right directions." Using the "think strategy" that I taught my students, I managed to stay in refueling position for three minutes of actual instrument contact. Breaking out of the clouds, the boom operator congratulated me for a fine refueling contact. I returned the favor to him, also. It was a dangerous thing to undertake but it proved to me that under wartime conditions, I could have taken all my fuel and reached my target.

Chapter 39

TURBINE ENGINE BLADE FAILURE (B-47)

During the period that I was stationed at Plattsburgh, there was occasionally a time when higher headquarters would direct a surprise Emergency War practice exercise. This consisted of various ground exercises and tests as a prelude to an actual flight with four thermonuclear bombs on board.

Everything went fine through the ground exercises and the actual "alert" scramble and takeoff with a climb to 25,000 feet. After leveling off, I noticed a vibration in the plane. Checking around, I found the vibration in the number 4 engine throttle. All engines were normal. Retarding that throttle made the vibration almost disappear so it was surmised that the very high speed turbine bucket had failed in the engine. Safety dictated that the engine be shut down.

We continued the mission as planned and made a normal landing at the end of the exercise. We then turned the aircraft over to the crew chief and left for the debriefing. A few hours later I was called back to Operations and was asked to report back to my B-47 aircraft. The crew chief met me and proceeded to explain why I had been called out.

First, the aft turbine wheel, which is red hot at full throttle and spins at an extremely high rate of speed had lost a two inch segment of one of its turbine blades. The turbine bucket took a crazy path after failing. It ejected downward through the aluminum skin of the engine cowling, next hitting the hydraulic shock strut of the right outrigger landing gear, which, in flight, is retracted into the number four engine nacelle.

Secondly, the turbine blade made a left turn and penetrated the skin of the airplane surrounding the bomb bay. Then, passing through the maze of control wires it then hit the casing of a thermonuclear bomb and ricocheted through the bomb bay doors and out into space.

When this happened, it left holes in the engine cowling, the side of the plane and the bomb bay doors. There were gouge marks on the outrigger strut and the casing of the TN bomb. All the maintenance crews who examined the plane were quite amazed that the turbine blade had not caused irreparable damage to parts of the aircraft or, heaven forbid, penetrated the bomb and set off a dynamite explosion.

Chapter 40

Heart Attack

One day at a midwestern air base another pilot approached me and asked my opinion on his medical problem. This pilot stated that he knew of my repeated heart problem and wanted to compare his heart problem with mine. He went on to say that he had frequent heartburn after eating a meal and wondered if he could possible have a heart attack. I advised him that his problem was completely different than mine and that he should definitely see the Flight Surgeon. I left him with that advice and proceeded about my duties.

However, a week later he was scheduled for his first official flight. On that day he planned to take off with other crews on a simulated combat mission. But fate intervened and he was delayed for takeoff because of a maintenance problem. After this short delay, he got his crew together, a three-man crew, and proceeded to taxi out for take-off. In the meantime, the weather had worsened. Getting clearance from the tower, he rolled down the runway with the hope of catching up with his fellow pilots. However, nature intervened and just after takeoff, maybe 1,500 feet, he encountered the storm, spiraled out of the clouds for a sudden stop in a farmer's wheat field. All three crew members were killed instantly.

The honor system in the service was such that you didn't squeal on your buddies. If I had broken the "code" he and his crew would be alive today. A reported autopsy on the pilot's heart indicated that he had experienced heart attacks recently. His "heart burn" had probably been and actual heart attack. Having recently "checked out" in this type of airplane, he was so eager to fly that he felt seeing a doctor might "ground him" which would jeopardize his flying career. Should I, or anyone, have broken the "code" and reported him to the Flight Surgeon?

Chapter 41

SNOW PLOW

After spending two years teaching the B-47 at McConnell AFB, Wichita, Kansas, I was transferred to Lincoln AFB, Nebraska for two years. No unusual happenings occurred at either of these bases. We were then transferred to Mt. Home AFB, Idaho.

One thing about Mt. Home weather was that it was always perfect. However, we compensated the good weather by flying to Anchorage, Alaska for standby combat air duty. Two days in a row we flew our B-47s to Whitehorse, Canada with the intention of landing at Elmendorf AFB, Anchorage, Alaska. The first two trips, the weather went below weather minimums so we returned to Mt. Home.

The third trip was the charm, or so I thought. There was a twenty knot cross wind from the north with drifting snow. The runway was reported to have twelve inches of packed snow. No problem there, or so I thought. Descending from altitude, weighing more than maximum weight for landing wasn't my idea of a good landing configuration. The Elmendorf weather might close down if I waited too long and what with two planes behind me, I decided to land.

Approaching the airport was a breeze. As we landed, in an angled approach, all seemed well except that we didn't deploy the approach chute as it would make matters worse. Rolling down the runway, we deployed the brake chute and applied the brakes. I suddenly discovered that we were breaking down through the packed snow as large chunks of snow were flying in every direction. To top it off, there was a sheet of ice under the packed snow. This was suddenly a toboggan ride with no braking action and the end of the runway coming up fast! Expecting the inevitable, a crash off the end of the runway, I kicked hard left steering and still hitting the brakes, hoping to ground loop the airplane.

The miracle happened. As we reached the last 2,000 feet of runway, we hit a dry spot where the fighter jets run up their engines before takeoff. Instinctively, the plane made a hard left turn, which put us on the taxiway. What a lucky turn of events! Before rejoicing too much, I realized that we were still traveling pretty fast. Still applying full brakes, we didn't stop until reaching the hanger area. Only then did I thank God for providing me with a pathway to safety.

The fire crash trucks had been alerted for a possible crash and had followed us into the hanger area with the tower officer close behind. He said we had dug a deep ditch down the middle of the runway. All's well that ends well!

Chapter 42

COMPLACENT

Complacent is defined in the Readers Digest Family Word Finder as self secure, content, self-satisfied, smug, untroubled and at ease with the world. From my many flight safety lectures I've listened to or about, complacency should not exist in the cockpit. It is trouble waiting for a place to happen. That is what happened on a recent flight from Anchorage, Alaska to Mountain Home, Idaho at night with no moon.

Arriving over the Mountain Home VOR radio station, I was cleared for a jet penetration approach. This consisted of flying an outbound heading while descending two thirds of the altitude, then turning inbound and descending the other third, followed by a straight-in landing approach.

It seems simple enough until you think that the airplane will fly by itself. In my case, when I flew outbound the air was so smooth (forgot to put the landing gear down) that it put me in la-la-land. Suddenly, in the descent, I felt that something was wrong. I grabbed the elevator controls and pulled up as hard as I could, noting the altitude we were at. We had leveled off at fifty feet higher than the airport elevation. I had not made the proper turn inbound that was required. We almost plowed up a lot of sagebrush.

Climbing back to landing pattern altitude, we circled back and landed properly. Not only was this too close for comfort but it meant that I had become complacent and, therefore, I should quit flying. The next day I applied for retirement from the Air Force. Instead of letting me retire, the Air Force sent me to C-130s, a cargo-carrying airplane. Maybe that would be a safer place for me.

Chapter 43

LOOK OUT ABOVE

January, 1966 found me flying the C-130 at Dyess AFB, Abilene, Texas. Maybe I would be safer. Part of the C-130 training requires the crews to fly over a point on the bombing range and drop a dummy load, a pallet, on a given target. In order to determine the accuracy of their drop, a range officer positioned himself on the ground at a safe distance from the target but close enough so he could judge the accuracy of the drop. I was that range officer. When the plane called "cargo away" I looked up but couldn't see the 2,000-pound pallet on its way down with a parachute.

Suddenly, while looking up at the sky, I noticed that I couldn't see any stars directly overhead. It was headed straight for me. Performing a quick hundred-yard dash, I heard the pallet land behind me, where my lookout station was. But I didn't think that being a simulated pancake for the Air Force was my idea of having fun.

My retirement application was again submitted shortly thereafter.

Chapter 44

WARSAW

While stationed temporarily in England, I had the opportunity to fly as copilot on the embassy "food run" to Warsaw, Poland by way of Frankfort and Copenhagen. Leaving Copenhagen we had to make good on a "time block" to fly the corridor to Warsaw. The weather was fine. During the flight a Russian MIG followed us all the way in. He was flying a crossing pattern behind us, and keeping us continually in sight. Our C-130 had a camouflaged paint job on it for use in Vietnam.

The pilot-in-command had been notified before leaving England that he was on orders to report for duty in Vietnam in sixty days. He was furious. Before we left England he tried everything he could to get his orders changed but to no avail. That is why he let me do all the flying he was still upset about having to go to Vietnam.

At Warsaw we had three hours to unload our cargo, reload if necessary, fuel up and prepare for departure. While the crew was doing that we were given a quick tour of the west bank of the Wista River, which flows through the city of Warsaw.

Clearing security, we prepared for takeoff. There was one problem, however. The number two engine would not start and the maintenance officer who was supposed to go with us had inexplicitly gone on another C-130 to Italy. What to do? The pilot-in-command asked me if I would be willing to fly the plane back to Denmark on three engines. I said I would. Calling for taxi instructions, the lady in the tower, with a clipped English accent, gave me permission to taxi and takeoff. She added, however, that our number two engine was not turning. We acknowledged her message by stating that we didn't need that engine for a flight to Copenhagen. She acknowledged that and cleared us for takeoff. In actuality, if we did not take off on time to keep our block time clearance, we would have to wait in Warsaw for thirty days waiting for another clearance to leave. Besides, the pilot-in-command was chomping at the bit to get back to England to continue fighting for a release on his orders to go to Vietnam.

Executing our takeoff on time, we made a right turn after takeoff with the MIG again following closely behind. Arriving in Copenhagen, we made a safe landing and taxied to the parking area. After repairing our bad engine, we flew back to England the next day.

Chapter 45

LUCKY ME

Retiring from the Air Force in 1967 and returning to Wichita, Kansas from Alaska and a ferry flight to Australia, I looked for other worlds to conquer. With my experience flying the Twin Otter in Alaska, I was asked to ferry one to Nairobi, Kenya, Africa. It seems that the chief pilot of that particular airline in that country was supposed to proceed to Toronto, Canada and take delivery of their new plane, the Twin Otter. But before he could leave, he had a household chore that he had to take care of first. When last seen alive, he was drilling a hole in the bathroom wall and hit an electric power line. He was electrocuted on the spot.

Now someone else had to deliver the plane to Nairobi. Up steps your former bush pilot from Alaska. Securing all the necessary maps and paperwork to make the trip, I proceeded to the Air Terminal in Wichita. While walking the underground passageway to the airplane, I was suddenly struck with a peculiar "chill" followed by a short period of sweats. I wondered if this was a sign. But I made the trip anyway.

Arriving in Toronto and getting briefed on everything they wanted me to do, I proceeded to my hotel for the night. The next morning I woke up with the worst case of flu imaginable. After talking to a hotel doctor, he said that I had contracted the Hong Kong flu. After calling my office in Wichita, they chose a new pilot to take my place. He arrived the next day and all papers and documents I turned over to him.

For three days I laid in bed with a high fever and the sweats. The telephone operator was very kind to me and sent hot soup up to me three times a day. On the fourth day I returned to Wichita.

After Captain Smith returned from Nairobi he told me about his hair-raising trip. His route went via the Azores, Portugal, Spain, and then across the Mediterranean Sea to Benghazi, Africa. When he landed, he was almost out of fuel. All fuel gauges showed almost empty when he should have had a substantial reserve. This would be something to check out.

The next day when he was pre-flighting the aircraft he opened a lower hatch to check inside for leaks. A ton of fuel came pouring out of the hatch. It was obvious he had a leak. After letting all the fuel drain out, he discovered that a major fuel line clamp had not been tightened during manufacturing and was also missed by the final inspectors.

Not only was there a serious problem of running out of fuel over the Mediterranean, there was the additional danger of an explosion in the fuselage caused by all the fuel vapor. In my case, the good Lord knew what he was doing when he arranged for me to get the flu.

Chapter 46

THE INSTRUMENT THAT LIED

One of my duties as an FAA Inspector was to give instrument check rides in reciprocating engine aircraft. In other words, the older planes we used before the arrival of jets. A request was made of me to administer a check ride in the DC-6. The pilot of this plane was an old hand at flying this aircraft but needed an annual check ride to update his proficiency. Preparations for the flight, the inspection, checklists, and all normal checks were completed satisfactorily.

One part of the flight check was the Instrument Landing System (ILS) approach for landing. The pilot flew the airplane around the flight pattern and lined the airplane up for the ILS. I noticed that it seemed odd that the pilot had lined up on the approach so expertly, that the crosshairs on the ILS showed "exactly" on the glide slope.

As we flew down the glide slope the crosshairs stayed centered. This must be a truly remarkable pilot, what with winds and turbulence trying to upset his fine-looking approach. In a few minutes I thought he would have completed his check ride. But my visual check on the altimeter was telling me that we were getting well below the glide slope, even though the ILS showed us on course and glide slope. There were no red warning flags to indicate any deviation from a perfect course alignment. This could not be. Something was wrong.

Ordering the pilot to remove his "instrument hood" and execute a "missed approach" the pilot was flabbergasted to see that he was dangerously low on altitude. Needless to say, he failed to use the altimeter in his cross check on the instrument panel. After landing, I called the avionics inspector and related what had happened. He didn't think anything like this could happen. He immediately caught a plane and proceeded to the airport where the plane was based and started an investigation. It turned out that a small obscure part in the ILS had failed to work properly. This kept the red warning flags from operating correctly and, as a result, the pilot was unaware his altitude was dangerously low. If the plane had been flying under adverse weather conditions with no instrument cross check, they could have easily flown into a building or a hill.

Chapter 47

CHRISTMAS ISLAND

While working in Burbank, California, the Federal Aviation Administration assigned me to various jobs, checking flight-crew members, checking the loading of hazardous materials, working up the annual budget, and certifying new airlines. Not to be overlooked is the responsibility to put an airline out of business if they fly in an unsafe manner. Also included is the filing of enforcement actions against any pilot who breaks the rules of the FARs.

One of my unusual assignments was to fly with and approve an airline application to start up a passenger route between Honolulu and the Christmas Islands, flying DC-6 aircraft. Leaving Honolulu early in the morning, we made the long flight to the Christmas Islands, which are due south of Hawaii and are located near the equator. After landing, I inspected the airport and all of the operational area that this airline would use.

When the flight crew completed their work and I was satisfied that this would be a safe operation, we prepared for takeoff. Just as the sun was setting we took off and headed north for Hawaii and cruising altitude. An hour into flight all four engines suddenly quit. At that time I was walking down the aisle making my inspections. The airplane started yawing left and right as the engines took turn quitting. Yours truly made a quick dive for the most available seat.

After the pilot got all four engines running, I went up to the flight deck to find out what happened. It seems that the pilot had wanted to disconnect the switches for the autopilot, re trim the plane and then reengage the autopilot switches. There was only one problem: the switches he disengaged were the fuel mixture controls. This pilot was used to flying the DC-6 but failed to remember that the switches were reversed on all other DC-6s. This was a prototype DC-6, one of a kind.

Chapter 48

CATASTROPHY IN THE MAKING

In 1970 while working as a safety inspector at McCarran Airport, Las Vegas, Nevada, I was privy to a conversation, told second hand, between a tower operator and the pilot of a scheduled airliner, a four engine plane.

Pilot: We have the airport in sight.

Tower: Cleared to land.

Pilot (On interphone to copilot): I can no longer see the lights of the airport, nor can I see the lights of Las Vegas.

Copilot (On interphone): Roger, on that. I can't see them either.

Pilot: Something is wrong; I'm pulling up. (Sound of increased power)

Conversation after landing:

Copilot: What do you think happened?

Pilot: Let's check the radios.

Copilot: Here it is. I set the Boulder VOR instead of the Las Vegas VOR.

Pilot: That could have been a wet landing.

Copilot: We almost landed in Lake Mead.

Chapter 49

WHOA NELLIE

On another flight in an L-188, I found a new way for a pilot to land an airplane. On arrival at Lake Havasu, flying east with directions from the tower, I noticed that we were traveling very fast over the lake. Informing the pilot was futile. He responded that this was a normal approach.

We ballooned down to the mid point of the runway before landing. After that, the pilot went into full reverse thrust on the propellers and full brakes. After parking the plane near the tower, I exited the plane and was relieved and concerned to feel a strong west wind. We had landed downwind.

Talking to the local weather observer, he showed me the wind direction was from the east at twenty miles per hour. I took him outside the building and asked him again, "Which way is the wind blowing from?" It wasn't until then that he noticed there was a strong wind from the west.

How could this be? We both looked at the wind arrow on the roof of the building. It was showing east. Upon closer inspection it was found that the weather vane was stuck in the east direction. How many planes had landed with a tail wind direction, and no pilot or weather observer had noticed this discrepancy?

Chapter 50

LESSON IN NAVIGATION AND TERRAIN CLEARANCE

One summer day I was flying from Los Angeles to Cleveland to supervise the passenger loading of charter passengers who were to be flown to Lake Havasu, Arizona. This flight into Havasu was the forerunner of the property development near the newly located London Bridge.

While flying the return route from Detroit, Michigan through an area south of Denver, Colorado, the pilot contacted Denver Center for a lower altitude because of a failing air pressurization system. We were at 20,000 feet with a cabin pressurization of 10,000. Denver cleared his plane to 16,000 feet. There was no minimum safe altitude listed on the charts. I informed the pilot of his mistake and that, further, we were about to encounter a mountain.

As the FAA Inspector, I informed the captain that getting a direct route versus airways was questionable. I had been through this area before and remembered that there were mountains in the area. At the present time we were in the clouds and could not visually navigate.

The captain maintained that if Denver Center had "cleared" him through the area, there simply could not be any danger of running into a mountain. I disagreed with him and informed him that human beings make mistakes. We continued our descent to 16,000 feet and then broke out of the clouds with good visibility. Dead ahead were some very large mountains. The captain broke out in a heavy sweat. We deviated around the mountain and proceeded on course. The passengers enjoyed a beautiful view of the mountains as we flew by them. That was a close call!

The captain was sanctioned for flying into a non-airways area. They also failed to use proper maps. The Denver Center was also criticized for allowing a plane to fly this unapproved route.

APPENDIX

1. P-43

2. P-22

3. BT-14

4. B-24

5. Westover Radio Range

6. Group Assembly

7. Collision Formation

8. AT-6

9. Lightning Strike

10. B-36

11. B-47 Nose Cone

12. B-58

13. B-47 Inflight Refueling

14. B-29

15. Exit

APPENDIX - PAGE 1

P-43 PHOTO (USAFM)

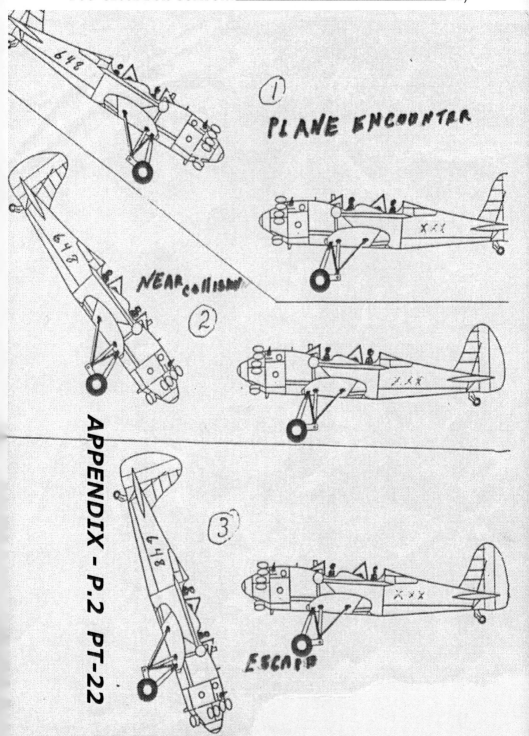

① PLANE ENCOUNTER

NEAR COLLISION

②

APPENDIX - P.2 PT-22

③

ESCAPE

PHOTO (USAFM) BT - 14B - 24

B - 24
PHOTO: (USAFM)

TWO
HOURS

PT "C" ←
(ACTUAL)

X, ← RANGE STATION   ■ BOSTON
O WESTOVER AFB
■ SPRINGFIELD   PT "D" (EST)

CAPE COD

■ HARTFORD   PROVIDENCE

ONE
HOUR

■ NEW HAVEN

LONG ISLAND

NEW
YORK ■

(NAVIGATION
AREA)

PHILADELPHIA
■

ATLANTIC OCEAN

WESTOVER RANGE
APPENDIX - P.5

A

CONSTANT EQUAL IDENTIFIE

BI-SIGNAL ZONES

ONLY ONE IDENTIFIER
QUADRANT UNKNO

N

RANGE STATION
N

A

WESTOVER FIELD
RUNWAY

DETAIL PLAN
of
GROUP ASSEMBLY

LEAD, HIGH and LOW SQUADRONS
FORM at 8,000, 9,000 and 7,000
RESPECTIVELY

2nd COMBAT WING

Buncher No. 6

Lead Group - Reserve

High Group - RA - 1000'
Low Group - RA - 1000'

Altitude

BOMBERS CLIMBING
TO FORMATION ALTITUDE
UNDER CIRCLE

OTHER GROUP

4573

5 MILES

5 MILES

5 MILES

Nº 6

BUNCHER

SQUADRON
GAINING
FORMATION

TO DIVISION ASSEMBLY

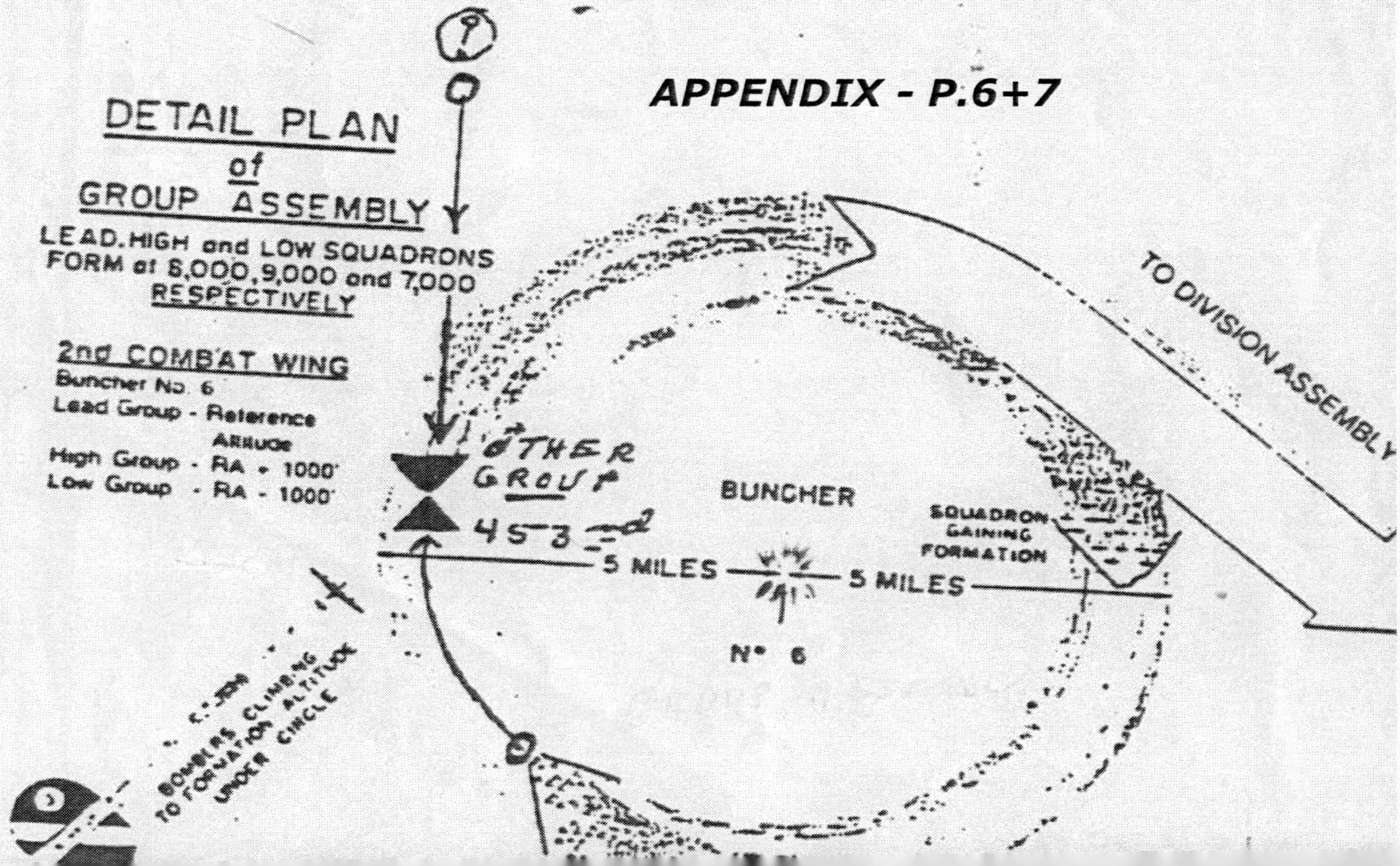

APPENDIX P.7

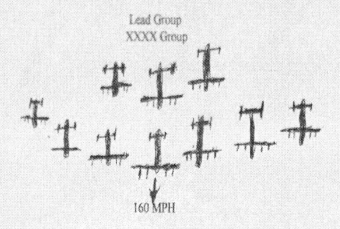

Lead Group
XXXX Group

160 MPH

A Case of Chicken
(Both groups at same altitude)

160 MPH

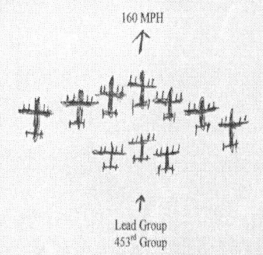

Lead Group
453rd Group

COLLISION FORMATION

APPENDIX P.8

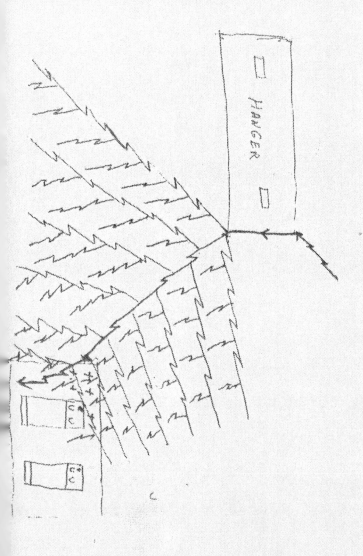

B - 36
PHOTO: (USAFM)

APPENDIX - P.11
B - 47 AND NOSE CONE
PHOTO: (USAFM)

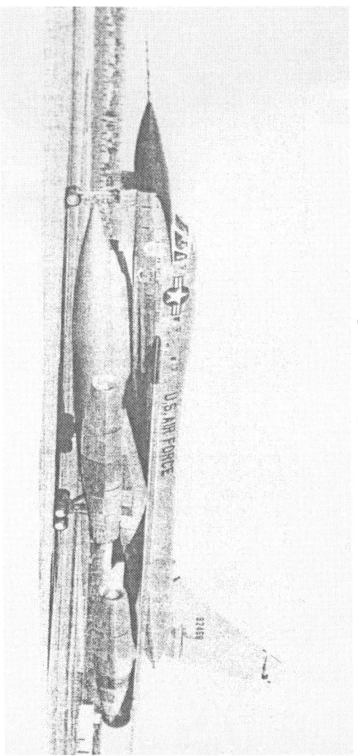

APPENDIX - P.11
B - 58
PHOTO: (USAFM)

COMBAT PHOTO

(453) 6 X11-10-44X101-10-00-00XForMATioN IPicTUR

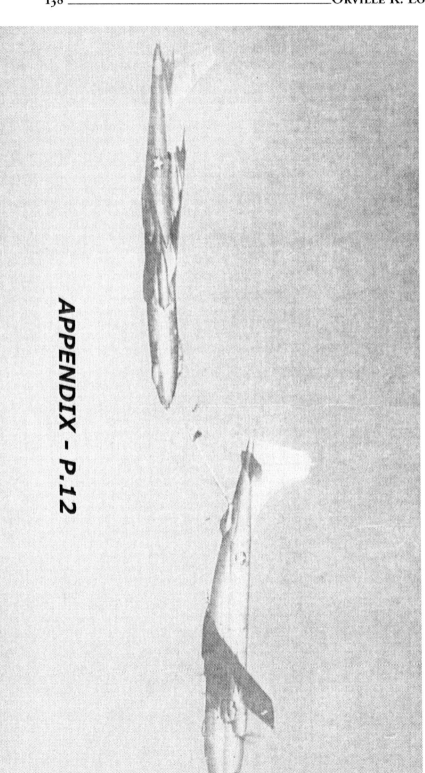

B - 47 IN FLIGHT REFUELING WITH KC - 97 TANKER

APPENDIX - P.12

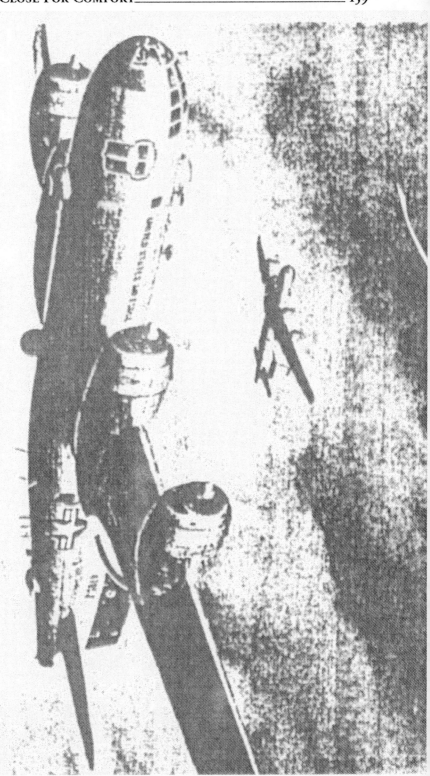

APPENDIX - P.12
B - 29
PHOTO: (USAFM)

Printed in the United States
1328000001B/118-147